U.S. Department
of Transportation

National Highway
Traffic Safety
Administration

DOT HS 809 729

May 2004

Technical Report

Safety Belt Use in 2003 – Demographic Characteristics

NCSA
National Center for Statistics and Analysis

1. Report No. DOT HS 809 729	2. Government Accession No.	3. Recipient's Catalog No.
4. Title and Subtitle Safety Belt Use in 2003 – Demographic Characteristics		5. Report Date May 2004
		6. Performing Organization Code NPO-101
7. Author(s) Glassbrenner, Donna, Ph.D.		8. Performing Organization Report No.
9. Performing Organization Name and Address Mathematical Analysis Division, National Center for Statistics and Analysis National Highway Traffic Safety Administration U.S. Department of Transportation NPO-101, 400 Seventh Street, S.W. Washington, D.C. 20590		10. Work Unit No. (TRAIS)
		11. Contract or Grant No.
12. Sponsoring Agency Name and Address Mathematical Analysis Division, National Center for Statistics and Analysis National Highway Traffic Safety Administration U.S. Department of Transportation NPO-101, 400 Seventh Street, S.W. Washington, D.C. 20590		13. Type of Report and Period Covered NHTSA Technical Report
		14. Sponsoring Agency Code
15. Supplementary Notes		

Abstract

This report presents results on the demographics of safety belt use from the 2003 National Occupant Protection Use Survey (NOPUS), with particular emphasis on results that evaluate aspects of the 2003 Click It or Ticket campaign to raise safety belt use nationwide. NOPUS provides the only probability-based observational results on belt use on the road in the United States, and is conducted annually by the National Center for Statistics and Analysis in the National Highway Traffic Safety Administration (NHTSA).

The principal findings of the 2003 survey include that safety belt use increased substantially among those the advertising component of the 2003 campaign sought to reach, namely, males ranging between young adults and adults. The 2003 NOPUS also found increased use in urban and suburban areas, and among females in the same young-adult-to-adult age range. The latter may be because females were similarly affected by the advertising, or because they might be substantially influenced by the other major component of the 2003 campaign, highly visible enforcement activities conducted by police.

17. Key Words safety belt use, NOPUS, demographics, Click It or Ticket campaign	18. Distribution Statement Document is available to the public through the National Technical Information Service, Springfield, VA 22161 http//:www.ntis.gov		
19. Security Classif. (of this report) Unclassified	20. Security Classif. (of this page) Unclassified	21. No. of Pages 54	22. Price

Form DOT F 1700.7 (8-72)   Reproduction of completed page authorized

TABLE of CONTENTS

1. Executive Summary 1

2. Background: The 2003 Campaign to Raise Safety Belt Use 2

3. Highlights of the 2003 Survey 3

    3.1   Use Increases Among Young Adults and Adults, Both Males and Females 3

    3.2   Safety Belt Use Increases in Urban and Suburban Areas 6

    3.2   Safety Belt Use Increases in Urban and Suburban Areas 7

    3.3   Restraint Use Among Children, Ages 0-7 9

4. Methodology 10

    4.1   Conversion Rates 10

    4.2   Survey Design 12

        4.2.1   The NOPUS Design 12

        4.2.2   Changes in 2003 13

5. Additional Tables 14

6. References 23

TABLE of FIGURES

Table 1: Safety Belt Use by Gender and Estimated Age 5

Figure 1: Safety Belt Use Among Ages 8 and Up, By Gender 6

Table 2: Safety Belt Use by Urbanization 7

Figure 2: Safety Belt Use by Urbanization 7

Table 3: Restraint Use by Front Seat Children, Ages 0-7 9

Figure 3: The NOPUS Regions 15

Table 4: Sample Sizes 15

Table 5: Safety Belt Use by Various Characteristics 16

Table 6: Safety Belt Use by Gender and Vehicle Type 16

Table 7: Safety Belt Use by Urbanization and Vehicle Type 17

Table 8: Safety Belt Use by Gender and Urbanization 18

Table 9: Safety Belt Use by Gender and Geographic Region 18

Table 10: Safety Belt Use by Gender and Ambient Enforcement Law 19

Table 11: Safety Belt Use by Age and Urbanization 20

Table 12: Safety Belt Use by Age and Geographic Region 21

Table 13: Safety Belt Use by Age and Ambient Enforcement Law 22

1. Executive Summary

In May of 2003, the National Highway Traffic Safety Administration (NHTSA), in conjunction with highway safety offices in nearly every State of the U.S., conducted the largest ever campaign to increase safety belt use. Various evaluations of this campaign, which used the messaging theme "Click It or Ticket", found the campaign highly successful. Perhaps most strikingly, use nationwide rose an unprecedented four percentage points in a single year, from 75% in 2002 to 79% in 2003. (Solomon et al., 2003; Glassbrenner, 2003)

In addition to highly visible enforcement activities, a major prong of the effort to raise use was a $24 million nationwide advertising campaign that focused on the demographic of 18-34 year old males. In this report we add to the evaluations of the 2003 campaign by reporting on observed usage rates by demographic characteristics, such as age and gender, that particularly address the advertising component. These rates were obtained from the National Occupant Protection Use Survey (NOPUS), which is conducted annually by the National Center for Statistics and Analysis in NHTSA and is the only probability-based survey that observes belt use on the road in the United States.

Our main findings are:

- Safety belt use rose among 16-24 year old males and 25-69 year old males, indicating that the 2003 advertising campaign was a success. The campaign, whether through its advertisement or enforcement components, also appeared to reach females in the same age ranges.

- Despite the success in reaching males, use remains statistically lower among males than females.

- Belt use increased in urban and suburban areas of the country in 2003.

These findings supplement results reported in (Glassbrenner, September 2003), which found increases in safety belt use in the South, in all vehicle types, in all times of day and week, and in States governed by "secondary" belt laws. The results in (Glassbrenner, September 2003) are also from the NOPUS survey.

In this paper, sport utility vehicles may be referred to as SUVs.

2. Background: The 2003 Campaign to Raise Safety Belt Use

In May of 2003, NHTSA and highway safety offices in nearly every State conducted the largest ever campaign to raise safety belt use in the U.S., called the 2003 Click It or Ticket Campaign. The campaign involved two components, the combination of which has been shown effective in raising use rates – publicity and highly visible enforcement. Law enforcement officers in 43 States, the District of Columbia, and Puerto Rico wrote 500,000 tickets for nonuse. NHTSA purchased $8 million advertising on television and radio, and States purchased an additional $16 million. (Solomon et al., 2003; Tyson, 2003).

The 2003 campaign was much larger than previous campaigns. The next largest campaign occurred in the previous year, in which 250,000 tickets were reported to NHTSA and $5 million was spent on media. For more information on the 2002 and 2003 campaigns, see (Solomon et al, 2002; Solomon et al., 2003).

The Click It or Ticket advertising in 2003 was designed to reach 18-34 year old males, a demographic chosen by NHTSA for various reasons. Safety belt use is lower among males than females, and among younger adults than older adults, as NOPUS consistently finds. Also, young males generally engage in riskier behaviors (Shinar et al., 2001), and this demographic can be effectively reached by advertising during certain types of shows, such as sports and news programs.

The theme of the advertising campaign was "Click It or Ticket", conveying the message that it is illegal not to use safety belts, law enforcement officers are looking for nonuse, and that you will be ticketed for nonuse. Both advertising purchased by NHTSA and the States conveyed this message. These purchased advertisements, whether on radio, television, or in print, ran for two weeks, during the period May 12 – May 26, 2003. Television ads were shown largely during sports and news programs, such as NASCAR's Coca Cola 600, ESPN's Sportscenter, and during the NHL Conference Playoffs, but also during other shows, such as Saturday Night Live, viewed by a substantial percentage of 18-34 year old males. For information on the content of the advertisements, including scripts from television and radio advertisements, and a more extensive list of shows during which ads were shown, see (Solomon et al., 2003).

In addition to advertising purchased by NHTSA and the States, the campaign generated a substantial amount of "earned media," that is, articles and stories written in response to the campaign. For instance, a number of stories on local enforcement activities, such as checkpoints at which police would be issuing tickets for safety belt use violations, appeared in local newspapers, and on local radio and television news programs. The earned media was generally about the campaign or particular activities, and so, unlike the purchased media, did not generally focus on any particular demographic.

3. Highlights of the 2003 Survey

This section highlights key findings from the 2003 survey. Additional results may be found in Section 5.

3.1 Use Increases Among Young Adults and Adults, Both Males and Females

Because the advertising branch of the 2003 Click It or Ticket campaign was designed to reach 18-34 year old males, NHTSA was particularly interested in determining whether safety belt use increased in this demographic. Consequently NOPUS, which had generally collected demographic data in even-numbered years, was supplemented to collect age and gender data in 2003. Due to limited funding, the 2003 NOPUS did not collect other information that it usually collects on its demographic survey, such as belt use by race and the use of child restraints.

To be consistent with prior NOPUS surveys, the 2003 survey categorized ages into the same adult age groups, namely 16-24, 25-69, and 70+, used in past surveys, rather than redefining the age groups to include an 18-34 year old category. Had NOPUS used an 18-34 age group in 2003, it would not have been able to see whether use increased in this group. Also since data collectors determine the age category into which an observed motorist falls subjectively, as opposed to interviewing the motorist, an 18-34 year old category in NOPUS would not have captured the campaign's target demographic precisely anyway.

The survey indicates that the advertising campaign was a success. Safety belt use rose among 16-24 year old males from 65% in 2002 to 72% in 2003, and among 25-69 year old males from 73% to 78%. Use also rose among males overall, from 72% in 2002 to 77% in 2003. Each of these increases is statistically significant with 95% confidence. Of course, we cannot isolate the reason for these increases, which may have occurred because of the advertisements, the enforcement campaigns, some combination of these, or other factors.

Interestingly, the survey also found increases in use among females in the same age ranges. Use rose among 16-24 year old females from 73% in 2002 to 80% in 2003, and in 25-69 year old females from 80% to 85%. Perhaps females in these age ranges decided to buckle more often in 2003 in response to the Click It or Ticket advertisements, although the advertising was not tailored to them specifically. It could also be that females in these age ranges used belts more in 2003 as a result of the enforcement campaigns, or for some other reason.

In fact the rise in use among males and females in general was such that use remains statistically higher among females than males, a pattern that NOPUS has seen for years. The male-female gap remains at 7 percentage points. That is, in 2003 we have the same pattern of use, with females remaining at a statistically higher level than males, but with both genders at higher use levels than where they were in 2002.

Although both males and females showed the same five percentage point increase in observed use, the data are mixed as to which gender made the greater improvement. On the one hand, the five-point jump was statistically significant (with 99% confidence) for males, but not significant (only 58% confidence) for females. That is, based on the five-point jumps we saw from a sample of sites in 2002 and 2003, we are 99% confident that safety belt use rose among males, but only 58% confident that use rose among females.

The level of confidence reflects the design of the NOPUS sample and the numbers of motorists observed, as well as the actual belt use rates in 2002 and 2003. The substantially lower confidence in an increase in female use could indicate that males improved their use rates, while females might have stayed the same. However note that Table 1 indicates a (statistically insignificant) decline in use among 8-15 year old females, and this may have contributed to the lower level of female significance. Also the lower female confidence could simply reflect that fewer females were observed than males in both survey years. About 19,000 males were observed in 2003 and 27,000 in 2002, compared to 16,000 females in 2003 and 23,000 in 2002. (The 2003 survey observed more motorists than the 2002 survey because it collected data during a greater total amount of time. See Section 4.2 for more information.)

On the other hand the five-point jump in each gender indicates that females made greater strides than males because the females started at a higher use rate in 2002. Nonuse among males decreased by 18%, from 29% in 2002 to 23% in 2003, while that for females fell 24%. Although the NOPUS use estimates reflect a "snapshot" of use on the roads, and not the percentage of the male and female populations that buckle to some specified degree, these reductions in nonuse indicate that roughly a quarter of female nonusers in 2002 converted to users in 2003, compared to 18% for males. (We do not mean to suggest by our choice of words that these "conversions" reflect permanent changes in behavior.) We explain this example in greater detail in the section "Conversion Rates".

Table 1: Safety Belt Use by Gender and Estimated Age

Motorist Group	Use in 2003		Use in 2002		2002-2003 Change			
	Estimate[1]	Standard Error	Estimate[1]	Standard Error	Estimate[2]	Standard Error	Degree of Confidence that Use Changed[3]	Conversion Rate[4]
Males	77%	2%	72%	1%	5% (S)	2%	99%	18%
ages 8-15	84%	4%	78%	5%	6%	6%	65%	27%
ages 16-24	72%	3%	65%	2%	7% (S)	3%	96%	20%
ages 25-69	78%	2%	73%	1%	5% (S)	2%	99%	19%
ages 70+	83%	4%	80%	2%	3%	4%	50%	15%
Females	84%	2%	79%	1%	5%	6%	58%	24%
ages 8-15	77%	13%	84%	3%	-7%	14%	40%	-44%
ages 16-24	80%	3%	73%	2%	7% (S)	4%	95%	26%
ages 25-69	85%	2%	80%	1%	5% (S)	2%	99%	25%
ages 70+	83%	8%	82%	2%	1%	8%	11%	6%

[1] An "H" or "L" indicates use that is statistically high or low in its category.

[2] An "S" indicates that use is statistically higher in 2003 with 90% confidence.

[3] The degree of statistical confidence in the assertion that use in 2003 was different than use in 2002.

[4] The conversion rate is the percentage reduction in safety belt nonuse.

Source: National Center for Statistics and Analysis, NHTSA, National Occupant Protection Use Survey

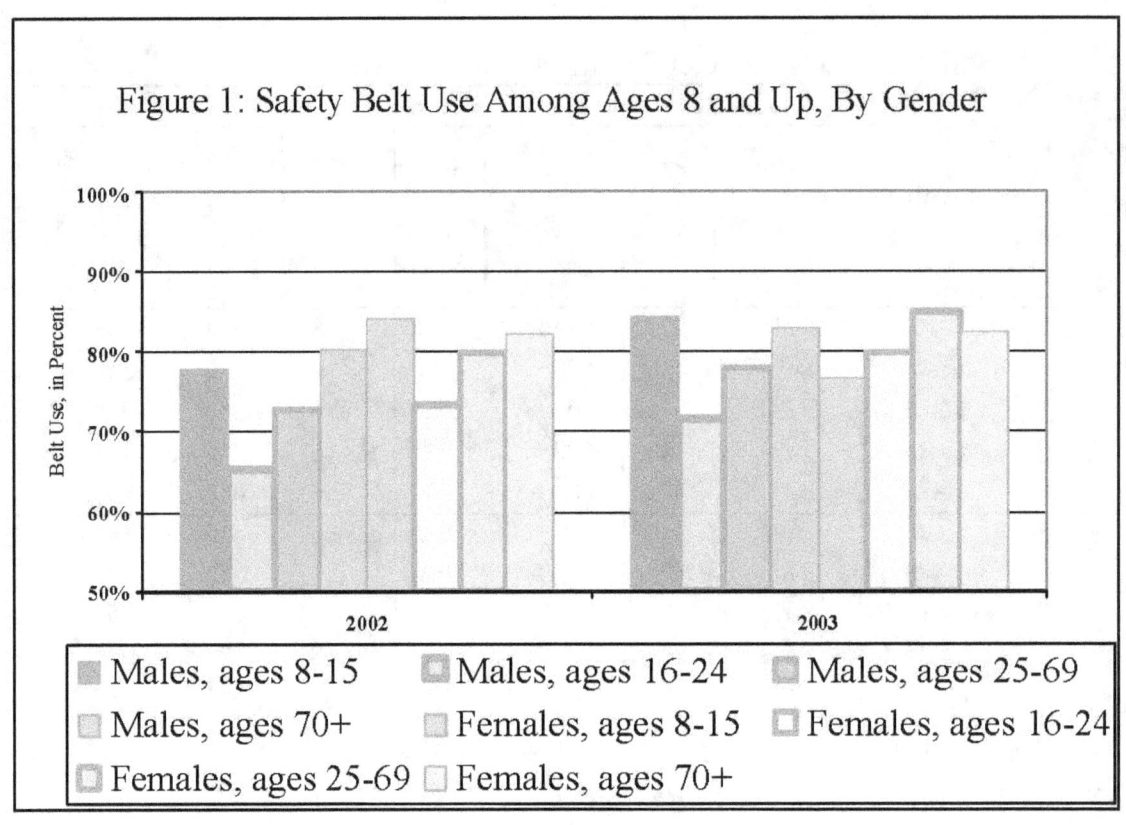

Figure 1: Safety Belt Use Among Ages 8 and Up, By Gender

Source: National Center for Statistics and Analysis, NHTSA, National
Occupant Protection Use Survey

3.2 Safety Belt Use Increases in Urban and Suburban Areas

Safety belt use in urban areas increased from 72% in 2002 to 79% in 2003, and in suburban areas from 76% to 81%. These are statistically significant with 90% confidence. One quarter of nonusers in urban areas were converted to users. (See Section 4.1 for an explanation of conversion rates.) Use in rural areas remained statistically unchanged at 74% in 2003.

Table 2: Safety Belt Use by Urbanization

Motorist Group	Use in 2003		Use in 2002		2002-2003 Change			
	Estimate[1]	Standard Error	Estimate[1]	Standard Error	Estimate[2]	Standard Error	Degree of Confidence that Use Changed[3]	Conversion Rate[4]
Urban motorists	79%	4%	72%	2%	7% (S)	4%	93%	25%
Suburban motorists	81%	2%	76%	1%	5% (S)	2%	96%	21%
Rural motorists	74%	3%	73%	2%	1%	3%	27%	4%

[1] An "H" or "L" indicates use that is statistically high or low in its category.

[2] An "S" indicates that use is statistically higher in 2003 with 90% confidence.

[3] The degree of statistical confidence in the assertion that use in 2003 was different than use in 2002.

[4] The conversion rate is the percentage reduction in safety belt nonuse.

Source: National Center for Statistics and Analysis, NHTSA, National Occupant Protection Use Survey

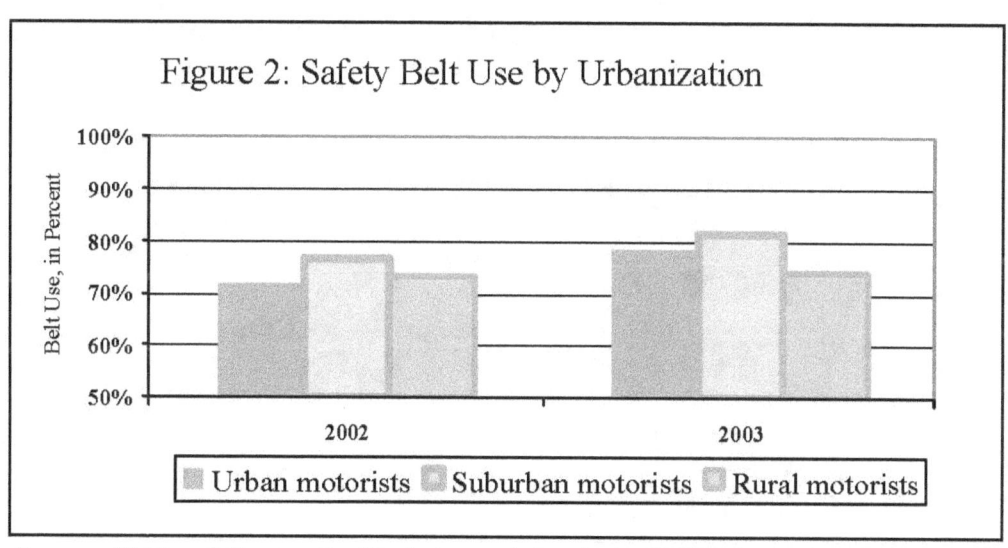

Figure 2: Safety Belt Use by Urbanization

Source: National Center for Statistics and Analysis, NHTSA, National Occupant
Protection Use Survey

In NOPUS, data collectors classify sites "Urban", "Suburban", or "Rural" subjectively as they visit each site. Urbanization is not assessed through independent means, such as by using data on population density. Consequently, the NOPUS urbanization categories tend to reflect the characteristics of the immediate area surrounding a site, as opposed to the population density of the city or town in which the site is located. For instance a developed downtown area of a sparsely –

populated town might well be classified as suburban or rural (although not likely "urban"). An advantage of using subjective classification is that it can capture changes in the sites' character quickly, while a disadvantage is inconsistency, since different data collectors may categorize the same site differently.

3.3   Restraint Use Among Children, Ages 0-7

NOPUS generally collects an extensive set of child restraint data in its demographic survey. NOPUS traditionally observes the use of various types of restraints used by children (forward-facing safety seat, rear-facing safety seat, booster seat, and safety belt) for various age ranges of children (e.g. ages 0, 1-3, and 4-7), in both front and rear seats of vehicles.   See (Glassbrenner, March 2003) for a detailed report on the 2000 and 2002 data.

However, due to limited funding, the 2003 survey observed only a <u>collapsed</u> age range of children (ages 0-7) in the <u>front</u> seat, and only observed whether these children were <u>restrained</u>, without noting the type of restraint used.   Children were counted as "restrained" if their shoulder belt was in use, whether it was used alone or in conjunction with some type of child seat (such as a booster seat).   Note that not all child safety seats use the vehicle's shoulder belt.   The findings of this observation are presented in Table 3.

In order to make comparisons between the 2002 and 2003 data, the 2002 data were recomputed for the restricted subpopulation of children observed in 2003.   That is, we took the data on front seat children in 2002, and collapsed the age ranges to a single 0-7 year old group.   Consequently the 2002 estimated use rate of 83% in Table 3 differs from the use rate of 88% for 0-7 year olds in (Glassbrenner, March 2003), which reflects <u>all</u> seating positions.

Table 3: Restraint Use by Front Seat Children, Ages 0-7

Motorist Group	Use in 2003		Use in 2002		2002-2003 Change				
	Estimate[1]	Standard Error	Estimate[1]	Standard Error	Estimate[2]	Standard Error	Degree of Confidence that Use Changed[3]	Conversion Rate[4]	
Children Ages 0-7	80%	7%	83%	4%	-3%	8%	30%	-18%	

[1] An "H" or "L" indicates use that is statistically high or low in its category.

[2] An "S" indicates that use is statistically higher in 2003 with 90% confidence.

[3] The degree of statistical confidence in the assertion that use in 2003 was different than use in 2002.

[4] The conversion rate is the percentage reduction in safety belt nonuse.

Source: National Center for Statistics and Analysis, NHTSA, National Occupant Protection Use Survey

NHTSA recommends that all children 12 and under should sit the back seat at all times, especially when the vehicle has active front passenger air bag(s). (Hurd, 2002; Hurd, 2004) We do not mean to suggest by Table 3 that the front seat is a safe seating position for children in this age range, even when they are restrained.

4. Methodology

4.1 Conversion Rates

Surveys that measure the percentage of a population that engages in a particular behavior, such as the use of safety belts, frequently evaluate improvement according to percentage point increases in the percentages of people who perform the behavior. According to this measure, males and females would appear to have shown similar improvements in safety belt use in 2002-2003, since both increased their use rates by five percentage points.

However, it was easier for males to increase five percentage points, because they needed to influence the behavior of a smaller portion of their nonusers to achieve this result. In 2002, 28% of males did not use safety belts, compared to 23% in 2003, an 18% reduction. That is, to increase their use by five percentage points, males changed the behavior of 18% of their nonusers. The corresponding percentage reduction in nonuse for females was 24%, i.e. females changed the behavior of nearly a quarter of their nonusers. That is, to increase use by five percentage points, males needed to change the behavior of a smaller fraction of their ill-behaving members because they started at a higher nonuse rate. (Here we are treating the NOPUS estimates as if they represented the percentage of males who buckle up to some degree. As explained in the next section, this is not true, but it illustrates the point.)

As we see in the previous paragraph, a fairer measure of improvement is the percentage reduction in nonuse, because it captures the amount of work that needs to be done to effect the change. We call the percentage reduction in nonuse the "conversion rate". That is, the conversion rate for males in the period 2002-2003 is 18%, while that for females in the same period was 24%. Note that the conversion rate is negative when use declines.

As we will see in the next section, the NOPUS estimates do not measure percentages of the population that buckled up, but rather a "snapshot" of use on the road. That is, if everyone froze at some (daylight) time in 2003, we would estimate that 77% of the male motorists (in a front outboard seat of a passenger vehicle) on the road at the time would be buckled up. Consequently, the 18% reduction in nonuse really means that there were 18% fewer unbelted males in the 2003 snapshot than in the 2002 snapshot. Thus thinking of the conversion rate as representing the percentage of the (in this case, male) population who were converted to using safety belts is not entirely accurate, but this interpretation provides a useful tool for assessing improvements in use and for comparing the relative improvements of different subpopulations, such as the accomplishments of males versus females.

It is important to keep in mind when considering conversion rates that these do not reflect permanent changes in behavior. That is, we think of 18% of males as being "converted" to using belts, but male belt use may well decline in the future.

Note also that conversion rates are based on measured increases in use, but we might or might not have statistical confidence that use increased at all. Recall from Section 3.1 that although the data indicated that both males and females experienced five-point jumps in use, statistically speaking, we are only confident in the assertion that males increased their use, having a relatively small degree of confidence (58% confidence) that female use increased. However, the calculation of a conversion rate is valid in each case. That is, we would still estimate that 24% of females were converted in 2003, even though we are only 58% confident that female use improved at all. Although this may seem contradictory, the first assertion represents our best measurement of a quantity (the percentage reduction in nonuse), while the second reflects our confidence in an assertion (the assertion that use increased). In a sense, the data give contrary indications on the increase in use among females.

In summary, the conversion rate reflects the percentage of nonusers whose behavior was modified. It is a fairer measure of the improvement in use rates than the percentage point increase in use because it reflects the amount of effort needed to achieve that increase.

4.2 Survey Design

4.2.1 The NOPUS Design

The National Occupant Protection Use Survey is the only probability-based observational survey of safety belt use on the nation's roads. Certain aspects of its design, some arising from practical restrictions imposed by its observational nature, are important to understand in order to properly interpret the NOPUS estimates.

In order to observe use as it actually occurs without influencing the results, NOPUS observes motorists on the road without stopping vehicles or interviewing motorists. Observations are made either from the roadside or from a moving vehicle. Due to practical restrictions on observing in these conditions, the survey restricts its belt use observations to observing the shoulder belt use of the driver and right front passenger during daylight hours (specifically, 8 AM – 6 PM). The substantial number of vehicles with tinted windows in the rear seat makes it challenging to observe rear seat use. In addition, it is difficult to observe the rear seat through the windshield, as would often be attempted at roadside vantage points, because the front seat often obstructs the view, especially in taller vehicles such as SUVs. However, we will investigate this further in the 2004 NOPUS. Consequently at this time the NOPUS estimates do not reflect lap belts, the rear seats, or nighttime use.

The NOPUS estimator produces "snapshots" of use. For instance, we would estimate that if everyone froze on the road in 2003 (during daytime), about 77% of males (in the front outboard seats) would be belted (with shoulder belts). Consequently, the NOPUS estimates do not represent the percentages of motorists who buckle up to some specified degree (e.g. at least half the time). The latter estimates are more frequently found in telephone surveys of use, such as (Block, 2001).

The NOPUS sample design uses a stratified cluster sample described in (Glassbrenner, 2002). Demographic data is collected at about 1,200 sites. See the same 2002 report for descriptions of variance calculations and for more information on the estimation formula that produces snapshots from data collected on vehicles passing a site.

Demographic data are collected at intersections that are controlled by a stop sign or stoplight, where slowed traffic gives data collectors more time to assess characteristics such as age and gender. This data is collected in what NOPUS calls its Controlled Intersection Study. There is reason to believe that belt use is higher at controlled intersections, since these occur more frequently in more populated areas, and NOPUS consistently finds higher (although not statistically higher) use in urban areas. Consequently NOPUS's demographic estimates are adjusted using data the survey collects at general intersections, from the NOPUS Moving Traffic Study, so as to remove this possible bias.

NOPUS was conducted during a 21-day period, from June 2, 2003 to June 22, 2003. This was shortly after the 2003 Click It or Ticket campaign ended. Research indicates that safety belt use rises sharply during and shortly after a campaign, and then drops off slightly, for a net gain in use. (Solomon et al., 1999) Consequently, the NOPUS use rates may reflect levels that were temporarily sustained. However, since the 2002 NOPUS was also conducted shortly after a campaign, the 2002-2003 changes in NOPUS estimates reflect actual annual changes, and not the larger increase one expects to see between just before and just after a campaign.

For additional detail on the NOPUS design, see (Glassbrenner, 2002) and (Glassbrenner, September 2003).

4.2.2 Changes in 2003

The demographic estimates were improved in 2003, in that they used independent traffic counts. Counts of the vehicles that pass the observation site in some relatively short time period, such as 5 or 10 minutes, are key elements of the estimation process in producing "snapshots" of use. See (Glassbrenner, 2002) for details. Prior to 2003, the NOPUS demographic estimates used traffic counts that were collected during a time in which nondemographic data (such as use by vehicle type and region of the country) was observed. In 2003, the data collectors conducted separate traffic counts when collecting demographic data, and the demographic estimates use these counts, which more accurately reflect traffic volume at the time of data collection.

The duration of the collection of demographic data was shortened in 2003 because of limited funding. Previous surveys observed demographic characteristics for 40 minutes per site, while the 2003 survey collected demographic data for 15-20 minutes. (Two-person teams observed data for 15 minutes, while single data collectors observed for 20 minutes to give them time to collect more data.) In the future, we intend to return to a 40-minute data collection period.

The NOPUS survey is in the process of phasing in additional improvements in data collection technology that have not yet affected the collection of demographic data. The NOPUS Moving Traffic Study is moving towards a paperless data collection through the use of specially programmed Personal Data Assistants (PDAs), and towards the collection of interstate data from vehicles traveling the interstate. For more information on these data collection methods, see (Glassbrenner, September 2003). For practical reasons, the NOPUS Controlled Intersection Study, from which the estimates in this report were derived, continue to use for the time being NOPUS' traditional data collection methods (collecting data on paper forms, and observing interstate traffic at exit ramps). We continue to investigate expanding these improved data collection methods to the greatest extent possible in the NOPUS.

5. Additional Tables

With NOPUS observing so many characteristics of motorists (e.g. age, gender, race) and their surroundings (e.g. vehicle type, urbanization, and geographic region), many cross tabulations are possible. We present below the one-way tables, along with several two-way tables that we find of intrinsic interest.

The tables in this paper contain estimated use rates with their associated standard errors. The standard error of an estimate is a measure of the possible error incurred by sampling. One expects the actual value being estimated to fall within twice the standard error of the estimated value. For instance, we would say with 95% confidence that the actual safety belt use in urban areas is in the range 71% - 87%.

These tables also include some categories for which use was observed in (Glassbrenner, September 2003), namely the ambient belt law, NOPUS regions, and the time of day and week. Safety belt laws are categorized as either 'primary", meaning that police may stop and ticket a motorist simply for belt nonuse, or "secondary", meaning that a motorist must be stopped for another infraction, such as an expired license tag before being ticketed for nonuse. In 2003, 20 States, the District of Columbia, and Puerto Rico had primary laws, while 29 States had secondary laws, and one State (New Hampshire) effectively had no safety belt law. (In New Hampshire, it is legal for motorists over the age of 18 to be unbelted.) In 2003, two states passed primary laws, namely Illinois and Delaware.

NOPUS divides the country into four regions as follows:

Northeast:	CT, MA, ME, NH, NJ, NY, PA, RI, VT
Midwest:	IA, IL, IN, KS, MI, MN, MO, ND, NE, OH, SD, WI
South:	AL, AR, DC, DE, FL, GA, KY, LA, MD, MS, NC, OK, SC, TN, TX, VA, WV
West:	AK, AZ, CA, CO, HI, ID, MT, NM, NV, OR, UT, WA, WY

Figure 3: The NOPUS Regions

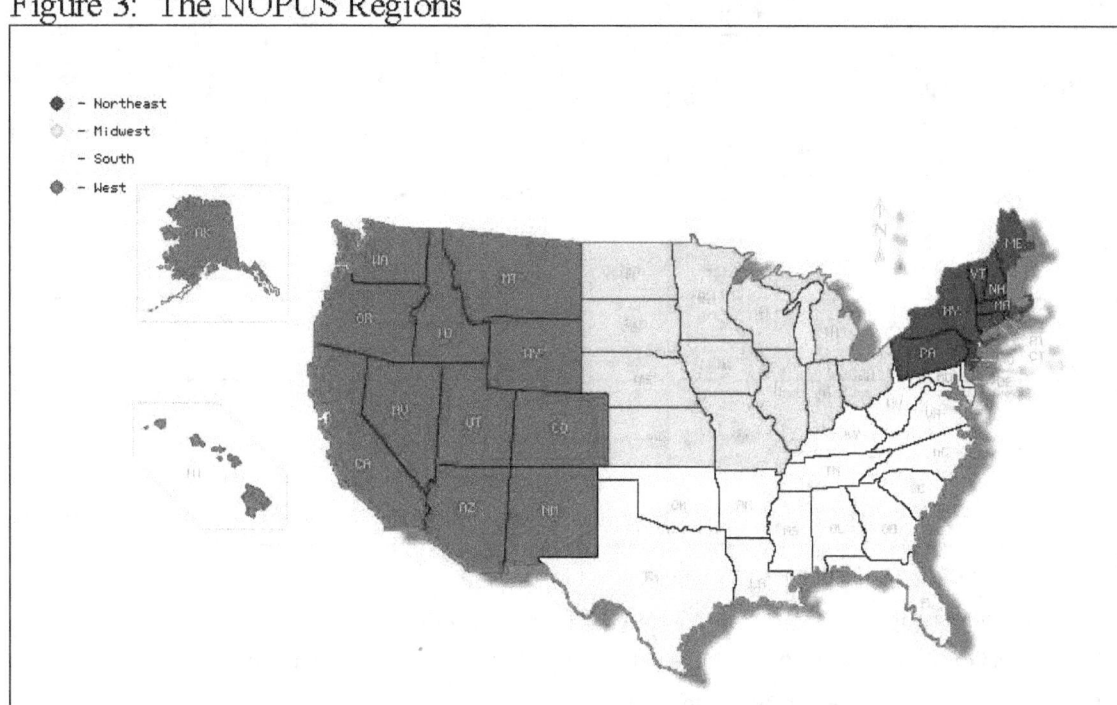

Legend:
- ● – Northeast
- ○ – Midwest
- – South
- ● – West

Source: National Center for Statistics and Analysis, NHTSA, National Occupant Protection Use Survey

Table 4: Sample Sizes

Numbers of	2003	2002
Sites	1,169	1,141
Vehicles	27,069	37,934
Front Seat Occupants	35,161	3,799
Ages 0-7	271	677
Ages 8-15	646	936
Ages 16-24	5,769	6,055
Ages 25-69	25,660	38,176
Ages 70+	2,815	4,330

Source: National Center for Statistics and Analysis, NHTSA, National Occupant Protection Use Survey

The 2003 NOPUS had fewer observations, because it observed at each site for about half of the time. Each site was observed for 40 minutes for the 2002 NOPUS and 15-20 minutes for the 2003 NOPUS.

Table 5: Safety Belt Use by Various Characteristics

Motorist Group	Use in 2003 Estimate[1]	Use in 2003 Standard Error	Use in 2002 Estimate[1]	Use in 2002 Standard Error	2002-2003 Change Estimate[2]	2002-2003 Change Standard Error	Degree of Confidence that Use Changed[3]	Conversion Rate[4]
Urban motorists	79%	4%	72%	2%	7% (S)	4%	93%	25%
Suburban motorists	81%	2%	76%	1%	5% (S)	2%	96%	21%
Rural motorists	74%	3%	73%	2%	1%	3%	27%	4%
Males	77%	2%	72%	1%	5% (S)	2%	99%	18%
Females	84% (H)	2%	79% (H)	1%	5%	6%	58%	24%
Ages 8-15	81%	6%	82%	3%	-1%	2%	40%	-6%
Ages 16-24	75%	2%	69%	2%	6% (S)	2%	99%	19%
Ages 25-69	80%	2%	76%	1%	4% (S)	2%	95%	17%
Ages 70+	81%	6%	82%	1%	-1%	3%	29%	-6%

[1] An "H" or "L" indicates use that is statistically high or low in its category.

[2] An "S" indicates that use is statistically higher in 2003 with 90% confidence.

[3] The degree of statistical confidence in the assertion that use in 2003 was different than use in 2002.

[4] The conversion rate is the percentage reduction in safety belt nonuse.

Source: National Center for Statistics and Analysis, NHTSA, National Occupant Protection Use Survey

Table 6: Safety Belt Use by Gender and Vehicle Type

Motorist Group	Use in 2003 Estimate[1]	Use in 2003 Standard Error	Use in 2002 Estimate[1]	Use in 2002 Standard Error	2002-2003 Change Estimate[2]	2002-2003 Change Standard Error	Degree of Confidence that Use Changed[3]	Conversion Rate[4]
Males	77%	2%	72%	1%	5% (S)	2%	99%	18%
in passenger cars	79%	2%	74%	2%	5% (S)	2%	98%	19%
in vans & SUVs	79%	3%	74%	2%	5% (S)	2%	99%	19%
in pickup trucks	66% (L)	3%	65% (L)	2%	1%	2%	33%	3%
Females	84%	2%	79%	1%	5%	6%	58%	24%
in passenger cars	84%	2%	80%	1%	4%	6%	50%	20%
in vans & SUVs	89%	2%	82%	1%	7%	4%	89%	39%
in pickup trucks	74%	6%	71% (L)	4%	3%	12%	19%	10%

[1] An "H" or "L" indicates use that is statistically high or low in its category.

[2] An "S" indicates that use is statistically higher in 2003 with 90% confidence.

[3] The degree of statistical confidence in the assertion that use in 2003 was different than use in 2002.

[4] The conversion rate is the percentage reduction in safety belt nonuse.

Source: National Center for Statistics and Analysis, NHTSA, National Occupant Protection Use Survey

Table 7: Safety Belt Use by Urbanization and Vehicle Type

Motorist Group	Use in 2003		Use in 2002		2002-2003 Change			
	Estimate[1]	Standard Error	Estimate[1]	Standard Error	Estimate[2]	Standard Error	Degree of Confidence that Use Changed[3]	Conversion Rate[4]
Urban motorists	79%	4%	72%	2%	7% (S)	4%	93%	25%
in passenger cars	81%	4%	72%	3%	9% (S)	4%	99%	32%
in vans & SUVs	82%	5%	72%	3%	10% (S)	6%	91%	36%
in pickup trucks	60% (L)	7%	69%	3%	-9%	7%	79%	-29%
Suburban motorists	81%	2%	76%	1%	5% (S)	2%	96%	21%
in passenger cars	83%	2%	78%	2%	5% (S)	2%	96%	23%
in vans & SUVs	85%	3%	79%	1%	6% (S)	3%	98%	29%
in pickup trucks	70% (L)	3%	69% (L)	2%	1%	3%	24%	3%
Rural motorists	74%	3%	73%	2%	1%	3%	27%	4%
in passenger cars	76%	4%	79%	1%	-3%	4%	50%	-14%
in vans & SUVs	83%	2%	78%	2%	5% (S)	2%	96%	23%
in pickup trucks	62% (L)	4%	54% (L)	5%	8% (S)	5%	92%	17%

[1] An "H" or "L" indicates use that is statistically high or low in its category.

[2] An "S" indicates that use is statistically higher in 2003 with 90% confidence.

[3] The degree of statistical confidence in the assertion that use in 2003 was different than use in 2002.

[4] The conversion rate is the percentage reduction in safety belt nonuse.

Source: National Center for Statistics and Analysis, NHTSA, National Occupant Protection Use Survey

Table 8: Safety Belt Use by Gender and Urbanization

Motorist Group	Use in 2003 Estimate[1]	Use in 2003 Standard Error	Use in 2002 Estimate[1]	Use in 2002 Standard Error	2002-2003 Change Estimate[2]	2002-2003 Change Standard Error	2002-2003 Change Degree of Confidence that Use Changed[3]	2002-2003 Change Conversion Rate[4]
Males	77%	2%	72%	1%	5% (S)	2%	99%	18%
in urban areas	76%	4%	67%	3%	9% (S)	4%	97%	27%
in suburban areas	78%	3%	73%	2%	5% (S)	3%	92%	19%
in rural areas	74%	3%	72%	2%	2%	4%	40%	7%
Females	84%	2%	79%	1%	5%	6%	58%	24%
in urban areas	85%	4%	78%	2%	7% (S)	4%	94%	32%
in suburban areas	86%	2%	81%	1%	5% (S)	2%	97%	26%
in rural areas	78%	4%	76%	4%	2%	5%	34%	8%

[1] An "H" or "L" indicates use that is statistically high or low in its category.

[2] An "S" indicates that use is statistically higher in 2003 with 90% confidence.

[3] The degree of statistical confidence in the assertion that use in 2003 was different than use in 2002.

[4] The conversion rate is the percentage reduction in safety belt nonuse.

Source: National Center for Statistics and Analysis, NHTSA, National Occupant Protection Use Survey

Table 9: Safety Belt Use by Gender and Geographic Region

Motorist Group	Use in 2003 Estimate[1]	Use in 2003 Standard Error	Use in 2002 Estimate[1]	Use in 2002 Standard Error	2002-2003 Change Estimate[2]	2002-2003 Change Standard Error	2002-2003 Change Degree of Confidence that Use Changed[3]	2002-2003 Change Conversion Rate[4]
Males	77%	2%	72%	1%	5% (S)	2%	99%	18%
in the Northeast	69%	3%	64%	2%	5%	4%	74%	14%
in the Midwest	71%	3%	69%	4%	2%	3%	46%	6%
in the South	79%	3%	73%	2%	6% (S)	3%	97%	22%
in the West	82%	5%	76%	3%	6%	5%	80%	25%
Females	84%	2%	79%	1%	5%	6%	58%	24%
in the Northeast	78%	2%	74%	2%	4%	3%	75%	15%
in the Midwest	80%	3%	80%	3%	0%	2%	0%	0%
in the South	89%	2%	80%	3%	9% (S)	2%	99%	45%
in the West	83%	9%	80%	2%	3%	7%	32%	15%

[1] An "H" or "L" indicates use that is statistically high or low in its category.

[2] An "S" indicates that use is statistically higher in 2003 with 90% confidence.

[3] The degree of statistical confidence in the assertion that use in 2003 was different than use in 2002.

[4] The conversion rate is the percentage reduction in safety belt nonuse.

Source: National Center for Statistics and Analysis, NHTSA, National Occupant Protection Use Survey

Table 10: Safety Belt Use by Gender and Ambient Enforcement Law

Motorist Group	Use in 2003		Use in 2002		2002-2003 Change			
	Estimate[1]	Standard Error	Estimate[1]	Standard Error	Estimate[2]	Standard Error	Degree of Confidence that Use Changed[3]	Conversion Rate[4]
Males	77%	2%	72%	1%	5% (S)	2%	99%	18%
under primary laws	80%	3%	78% (H)	2%	2%	2%	68%	9%
under secondary laws	73%	3%	64%	2%	9% (S)	3%	99%	25%
Females	84%	2%	79%	1%	5%	6%	58%	24%
under primary laws	86%	4%	83% (H)	2%	3%	4%	61%	18%
under secondary laws	82%	2%	74%	1%	8% (S)	2%	99%	31%

[1] An "H" or "L" indicates use that is statistically high or low in its category.

[2] An "S" indicates that use is statistically higher in 2003 with 90% confidence.

[3] The degree of statistical confidence in the assertion that use in 2003 was different than use in 2002.

[4] The conversion rate is the percentage reduction in safety belt nonuse.

Source: National Center for Statistics and Analysis, NHTSA, National Occupant Protection Use Survey

Table 11: Safety Belt Use by Age and Urbanization

Motorist Group	Use in 2003		Use in 2002		2002-2003 Change			
	Estimate[1]	Standard Error	Estimate[1]	Standard Error	Estimate[2]	Standard Error	Degree of Confidence that Use Changed[3]	Conversion Rate[4]
Ages 8-15	81%	6%	82%	3%	-1%	2%	40%	-6%
in urban areas	80%	9%	73%	10%	7%	15%	37%	26%
in suburban areas	83%	6%	83%	3%	0%	5%	0%	0%
in rural areas	74%	14%	82%	4%	-8%	14%	44%	-44%
Ages 16-24	75%	2%	69%	2%	6% (S)	2%	99%	19%
in urban areas	77%	4%	65%	4%	12% (S)	5%	98%	34%
in suburban areas	75%	4%	69%	2%	6%	4%	84%	19%
in rural areas	72%	3%	71%	3%	1%	4%	18%	3%
Ages 25-69	80%	2%	76%	1%	4% (S)	2%	95%	17%
in urban areas	79%	5%	72%	2%	7%	5%	87%	25%
in suburban areas	82%	2%	77%	2%	5% (S)	3%	95%	22%
in rural areas	76%	3%	73%	3%	3%	3%	66%	11%
Ages 70+	81%	6%	82%	1%	-1%	3%	29%	-6%
in urban areas	82%	9%	79%	4%	3%	10%	24%	14%
in suburban areas	87%	3%	82%	2%	5%	3%	86%	28%
in rural areas	60%	22%	84%	3%	-24%	21%	74%	-150%

[1] An "H" or "L" indicates use that is statistically high or low in its category.

[2] An "S" indicates that use is statistically higher in 2003 with 90% confidence.

[3] The degree of statistical confidence in the assertion that use in 2003 was different than use in 2002.

[4] The conversion rate is the percentage reduction in safety belt nonuse.

Source: National Center for Statistics and Analysis, NHTSA, National Occupant Protection Use Survey

Table 12: Safety Belt Use by Age and Geographic Region

Motorist Group	Use in 2003		Use in 2002		2002-2003 Change			
	Estimate[1]	Standard Error	Estimate[1]	Standard Error	Estimate[2]	Standard Error	Degree of Confidence that Use Changed[3]	Conversion Rate[4]
Ages 8-15	81%	6%	82%	3%	-1%	2%	40%	-6%
in the Northeast	88%	12%	93%	4%	-5%	18%	22%	-71%
in the Midwest	74%	14%	83%	6%	-9%	9%	70%	-53%
in the South	49%	12%	77%	17%	-28%	6%	99%	-122%
in the West	71%	15%	84%	7%	-13%	11%	78%	-81%
Ages 16-24	75%	2%	69%	2%	6% (S)	2%	99%	19%
in the Northeast	87%	4%	73%	7%	14% (S)	6%	99%	52%
in the Midwest	79%	7%	69%	4%	10%	7%	87%	32%
in the South	57%	9%	63%	3%	-6%	3%	98%	-16%
in the West	75%	8%	69%	3%	6%	7%	62%	19%
Ages 25-69	80%	2%	76%	1%	4% (S)	2%	95%	17%
in the Northeast	86%	7%	83%	2%	3%	4%	50%	18%
in the Midwest	84%	7%	79%	4%	5% (S)	2%	99%	24%
in the South	74%	7%	70%	4%	4% (S)	2%	97%	13%
in the West	82%	7%	78%	3%	4%	6%	51%	18%
Ages 70+	81%	6%	82%	1%	-1%	3%	29%	-6%
in the Northeast	91%	7%	88%	4%	3%	7%	35%	25%
in the Midwest	75%	22%	87%	1%	-12%	4%	99%	-92%
in the South	33%	18%	71%	7%	-38%	3%	99%	-131%
in the West	67%	21%	85%	1%	-18%	21%	61%	-120%

[1] An "H" or "L" indicates use that is statistically high or low in its category.

[2] An "S" indicates that use is statistically higher in 2003 with 90% confidence.

[3] The degree of statistical confidence in the assertion that use in 2003 was different than use in 2002.

[4] The conversion rate is the percentage reduction in safety belt nonuse.

Source: National Center for Statistics and Analysis, NHTSA, National Occupant Protection Use Survey

Table 13: Safety Belt Use by Age and Ambient Enforcement Law

Motorist Group	Use in 2003		Use in 2002		2002-2003 Change			
	Estimate[1]	Standard Error	Estimate[1]	Standard Error	Estimate[2]	Standard Error	Degree of Confidence that Use Changed[3]	Conversion Rate[4]
Ages 8-15	81%	6%	82%	3%	-1%	2%	40%	-6%
under primary laws	84%	10%	86%	3%	-2%	8%	19%	-14%
under secondary laws	77%	8%	76%	6%	1%	6%	13%	4%
Ages 16-24	75%	2%	69%	2%	6% (S)	2%	99%	19%
under primary laws	80% (H)	3%	73% (H)	2%	7% (S)	3%	96%	26%
under secondary laws	68%	5%	63%	3%	5%	4%	74%	14%
Ages 25-69	80%	2%	76%	1%	4% (S)	2%	95%	17%
under primary laws	82%	3%	80% (H)	0%	2%	2%	62%	10%
under secondary laws	77%	3%	69%	0%	8% (S)	2%	99%	26%
Ages 70+	81%	6%	82%	1%	-1%	3%	29%	-6%
under primary laws	78%	11%	87% (H)	0%	-9%	11%	59%	-69%
under secondary laws	85%	4%	74%	0%	11% (S)	3%	99%	42%

[1] An "H" or "L" indicates use that is statistically high or low in its category.

[2] An "S" indicates that use is statistically higher in 2003 with 90% confidence.

[3] The degree of statistical confidence in the assertion that use in 2003 was different than use in 2002.

[4] The conversion rate is the percentage reduction in safety belt nonuse.

Source: National Center for Statistics and Analysis, NHTSA, National Occupant Protection Use Survey

6. References

Blincoe, L., Seay, A., Zaloshnja, E., Miller, T., Romano, E., Luchter, S., Spicer, R., *The Economic Impact of Motor Vehicle Crashes, 2000*, NHTSA Technical Report, DOT HS 809 446, May 2002

Block, A., *2000 Motor Vehicle Occupant Safety Survey, Volume 2: Seat Belt Report*, NHTSA Technical Report, DOT HS 809 389, November 2001

Glassbrenner, D., *Safety Belt Use in 2003*, NHTSA Technical Report, DOT HS 809 646, September 2003

Glassbrenner, D., *Safety Belt Use in 2002 – Use Rates in the States and Territories*, NHTSA Research Note, DOT HS 809 587, May 2003

Glassbrenner, D., *Safety Belt Use in 2002 – Demographic Characteristics*, NHTSA Research Note, DOT HS 809 557, March 2003

Glassbrenner, D., *Safety Belt and Helmet Use in 2002 – Overall Results*, NHTSA Technical Report, DOT HS 809 500, September 2002

Hurd, T., *USDOT Requires Improved Child Restraint Labels*, NHTSA Press Release, NHTSA 63-02, October 2002

Hurd, T., *Survey Finds Widespread Misuse of Air Bag On-Off Switches in Pickups*, NHTSA Press Release, NHTSA 2-04, January 2004

Shinar, D., Schechtman, E., and Compton, R., *Self-Report of Safe Driving Behaviors in Relationship to Sex, Age, Education and Income in the U.S. Adult Driving Population*, Journal of Accident Analysis and Prevention, Vol. 33 (1), pp. 111-116, January 2001

Solomon, M., Nissen W., Preusser D., *Occupant Protection Special Traffic Enforcement Program Evaluation*, NHTSA Technical Report, DOT HS 808 884, April 1999

Solomon, M., Ulmer, R., Preusser D., *Evaluation of Click It or Ticket Model Programs*, NHTSA Technical Report, DOT HS 809 498, September 2002

Solomon, M., Chaudhary, N., Cosgrove, L., *May 2003 Click It or Ticket Safety Belt Mobilization Evaluation*, NHTSA Technical Report, no DOT number available, November 2003

Traffic Safety Facts 2002 – Occupant Protection, NHTSA Fact Sheet, DOT HS 809 610, undated

Tyson, R., *U.S. Transportation Secretary Mineta Launches Massive Law Enforcement Mobilization for Traffic Safety*, NHTSA Press Release, NHTSA 19-03, May 2003

DOT HS 809 729
May 2004

U.S. Department
of Transportation

**National Highway
Traffic Safety
Administration**

People Saving People
www.nhtsa.dot.gov

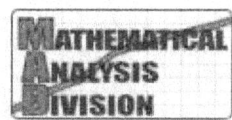